NIELS NEERSØ

A Viking Ship

"ROAR EGE"
– A reconstruction of a trading vessel from the Viking Age

PHOTOS: GRETHE SCHANTZ

BREAKWATER

Olaf Trygveson's saga

Torberg walked over to the ship and smoothed the gunwale with his axe so that all the disfigured parts were evened out. The king and all the others could see at once that the ship looked much better on the side where Torberg had cut than on the other one.

The king thanked Torberg and asked him to smooth the other side of the ship in the same way. The king also made Torberg foreman of the work on the ship until it was completed.

It was a dragon-ship, built on the same principles as "The Snake" but much bigger and in every detail made with greater care.

The king named it "The Long Snake" (Ormen den lange).

This is a little bit of what Snorre Sturlason's Icelandic Saga tells us about the Norwegian Viking-king, Olaf Trygveson, who reigned at the end of the eleventh century.

We can learn two things about the building of a Viking-ship from this saga.

First, that it was nescessary to have a foreman on the job, one man responsible. The Vikings called this man the "stemsmith." He knew – could

A detail from the Bayeux Tapestry, the dramatic scenes showing the Norman conquest of England. Duke William of Normandy orders many ships to be built, and this scene shows us the stemsmith at work. He is the man who is looking appraisingly at the ship at the top of the picture.

This is how beautifully correct it looked in Roskilde harbour while the "Roar Ege" was being built. Present-day Danes could follow the example given by their Viking forefathers.

see – what the ship should be like. The stemsmith knew in advance the shape and measurements of all the ship's details and how they should be placed in relationship to each other. He carried much of this knowledge in his head but for some measurements and angles he used a series of yard-sticks with notches.

Bow and stern were, however, decisive for the rest of the ship. The overall shape and every detail had to be absolutely exact with a tolerance of only milimetres. If this was done correctly, the rest of the ship could be adjusted accordingly.

Secondly, the name of the ship was the "Snake." It fit perfectly. The ships the Vikings

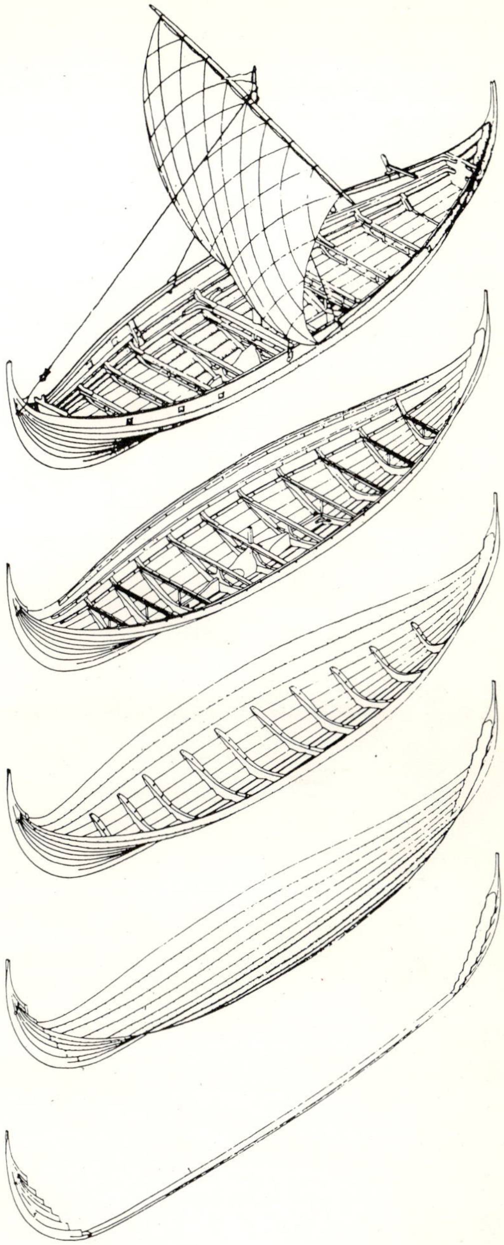

The way to work if you want to build a Viking-ship.

built and sailed so expertly really behaved like snakes in the water. The light, flexible hull gave way to – co-operated with – the movements of the water, where a modern steelship forces its way forward through the waves.

Viking ships were the technical wonders of that time. Centuries of experience in shipbuilding and sailing took the Vikings on plundering and trading expeditions to England, down the coast of Europe to the remotest corners of the Mediterranean, by Russian rivers to Byzantium and, probably, to America centuries before Columbus.

This book tells the story of how and why a small group of ordinary present-day Danes and some eminent scientists followed the example set by the old Vikings and built a true copy of a Viking ship, working under Viking Age conditions.

A beginning – and a dream

It all started with the discovery of Viking ships at Skuldelev: the unique archaeological event in 1957-62 when the remains of five Viking ships were found and removed from their resting place in the mud at the bottom of Roskilde fiord, where they had lain for nearly one thousand years.

It was well-known that something or other lay there in the Skuldelev Narrows. A longish elevation across the twisted channel contained big stones and ancient ships' timbers. The tale of Queen Margrethe's ship, which was believed to have sunk somewhere in the fiord, was commonly talked about.

When the local fishermen in the 1920s began using deep-draught motorboats they had to dig a passage through the middle of the obstruction across the narrows.

Among other things they found a more than four yards long keelson (a wood block where the bottom end of the mast is placed).

The artificial island within the interlocking steel-piles at Skuldelev in 1962. Wreck no. 3 – the small trading-vessel – has just been dug out from the mud.

The keelson was brought ashore and measured and photographed by a man from the Danish National Museum. But the wood had already begun to dry out and crack and for that reason it was never brought to the National Museum for further investigation. It remained private property until a cold winter's day during the Second World War, when it was burned in a stove in Copenhagen!

In short, nothing much had happened.

But in 1956 things began to move fast. Two sportdivers from Roskilde found a well-preserved

The bow of wreck no. 3 even at this early stage gave a clear impression of the astounding workmanship and sure sense of form of the Viking craftsmen.

rib of a ship. The National Museum organized a thorough investigation and it was established that the find was considerably older than Queen Margrethe (she died in 1412).

Diving and digging began, but on a very tight budget – primitive means indeed. A small dinghy was the most advanced piece of equipment the first summer. But when it was discovered that there were several ancient ships forming a barrier right across the narrow sailing channel, money began to flow more freely and better equipment was obtained. It was decided that the wrecks should be excavated and preserved for posterity.

In order to do this, an area of 1600 square metres had to be diked. In 1962 the whole site was enclosed within interlocking sheet-piles. Pumping began and soon an artificial island was created in the middle of Roskilde fiord.

Many tons of stones had to be removed before the contours of the wrecks began to be visible. But then it was obvious to everybody that the results exceeded even the wildest hopes. Five Viking ships, all of a different type, were uncovered.

The best preserved was "Wreck No. 3," a small trading vessel from the eleventh century, shorter and broader than the Viking warships, the awe-inspiring fast-sailing "Longships."

Twenty years of research – and a jig-saw puzzle

Five Viking ships, a unique find, had been dug up. Now the problem was what to do with them.

First of all, the hundreds of larger and thousands of smaller pieces of wood were photographed, measured and filed. Then everything was wrapped up in plastic in order to keep the wood damp. If this had not been done, the pieces would have shrinked, cracked and the smaller pieces eventually disappeared.

Many truck loads of plastic-wrapped, fragile ships timber were then sent to the Danish National Museum, where scientists subsequently found a method to preserve the pieces so that they could be assembled again in the original form.

The reconstruction of this three-dimensional jig-saw puzzle has been on-going for twenty years in the Viking Ship Museum, built in Roskilde in the years 1966 and '67. The five Skuldelev ships can now be seen in the museum, painstakingly reassembled piece by piece.

Skuldelev 3 - the small trading-vessel - in its place in the Museum. The reconstruction "the Roar Ege" will have its permanent anchorage just outside.

The Viking Ship Museum, seen from the fiord.

Plans and preparations

This brings us to the year 1981, when the dream of building an exact copy of the best preserved ship from the Skuldelev find suddenly became a tangible reality. A private foundation granted one million Danish kroner and on the basis of that the Viking Ship Museum could launch this unparalleled and ambitious project.

The building-team, a score of young men and women who were to do all the practical work under the guidance and control of the scientists from the Viking Ship Museum, were all volunteers. Most of them had some experience in building ships of wood; some had none. Some held university degrees; others were craftsmen. The whole team was determined to stay on to the bitter end, and the work was organized in such a way that a few of the team worked permanently while the remainder only turned up at weekends.

The task was difficult. For the first time in history they were to build and test the sailing ability of an exact copy of a real, well-preserved Viking ship as a scientific experiment. And do it while using only the tools which the Vikings themselves had used.

The scientific requirements were clear: this must be a professional, critical and analytical project placing equal importance on theory and the practical side.

For twenty years the scientists had studied, measured and calculated each and every detail in the construction of Viking ships, and how every part should be placed in relation to other parts. They had also examined the marks left by the Viking axes, chisels and augers on the Skuldelev ships. Now all measurements and calculations had to be put through the final test of practical reality.

The oak tree

The first big hurdle that had to be overcome was finding sufficient oak wood for the ship. Some of it had to be big, straight, completely flawless trunks.

The Vikings had enough to choose from, but the vast oak-forests that covered parts of Denmark in the Viking Age no longer exist. They were felled and used for housebuilding in medieval times and in later centuries in constructing the ships of the Royal Danish Navy.

Search parties were sent through the remaining forests in northern Zealand, where sufficient oak wood was found. The biggest trunk measured more than one metre in diameter. In addition to these big trunks, it was nescessary to find a large quantity of crooked trees and branches for "floors," "frames" and "knees," the terms boatbuilders use for some of the pieces of wood which brace the inside of a ship.

The oak trees for use in the "Roar Ege" are felled by hand in North Zealand.

This is what it looked like in the oak forests of Normandy in 1065, when Duke William had given the order "to build many ships." The Normans were mostly descendents of the Vikings who had conquered their country at the end of the Viking Age. It is therefore not at all strange that this picture shows us Viking ships. In the picture it is possible to recognize several of the characteristic axes, among them the broad-axe.

The tools of the Vikings

It was essential for the whole scientific character of the project that only Viking Age techniques and tools were used.

That was actually done. It is important to emphasize the fact that the whole construction of "Roar Ege" was completed without using a saw at all. The hard oak wood was hewn, chiselled and planed, where a present day builder of wooden ships would have used a saw.

Fortunately a great deal is known about the shipbuilding tools of the Vikings. On Gotland and at Hedeby archeologists have found a number of large and small axes, the impressive broad-axe, chisels, knives and several other tools as well.

It was also necessary to have a smith at the construction site at Roskilde harbour. It was there the iron spikes used in assembling the "Roar Ege" were made, while the building of the ship was going on.

A blacksmith at the National Museum making a big felling-axe by hand.

Drawings and tin-casts of these ancient tools were made and sent to the blacksmith at the National Museum, where copies were made by hand so that everything had the same weight and dimensions as the originals.

On the open beach

Now the oak wood was at hand, the tools had been made, the building-team had been assembled - and with all preparations finished, the Viking Ship Museum could begin the actual construction which took place in the open air shipbuilding yard in Roskilde harbour.

In the course of the early spring of 1983 several things were started at the same time.

The keel, a piece of oak nine metres long, was hewn in shape and firmly secured to piles that had been driven down into the slightly sloping ground of the building site. Bow and stern were also hewn from solid pieces of oak and these three parts, the keel, the bow and the stern, were joined together.

A simple lead hung on a line - a method the Vikings knew about.

The stern is given form.

At the same time, the builders began splitting the long oak trunks that were to become boards for the ship.

Boards is the shipbuilders term for the long, broad but comparatively thin planks forming the sides of a wooden ship. The splitting was done - as described in the caption - by an age-old method, the so-called radial cut technique. We know that the Vikings used it but today nobody has any experience with it at all. Consequently the builders had to experiment. Each trunk was a problem all by itself and for that reason alone some delays in the working schedule were unavoidable.

In order to obtain thin, but pliant and strong, wood for the boards, the building-team had to use an ancient technique, which is now forgotten. It is done in the following way: first, the long oak trunks are divided into two halves, these are then divided lengthwise into fourths, and then eights, and so on. As has already been pointed out, it was impossible to base the building of the ship on experience, and the building-team were gradually forced to recognize that the Vikings must have been better at this special technique than we moderns.

The keel of the ship is laid, and the first boards are in place. The bow and stern of the ship lift themselves elegantly towards the sky.

The builders, however, found it necessary to experiment in other ways too. One such was how to go about carding, spinning, twisting and impregnating the woollen threads that are laid in a narrow groove cut into the timber where two boards are joined in order to make the jointures watertight.

The tarred and twisted woollen thread used to make the jointures between the boards watertight.

It was also very difficult to hold the boards in exactly the curved shape and angles they had to keep in relation to one another. It proved nescessary to invent and make large, primitive wooden clamps and a system of supporting sticks. Both proved to be well suited to the purpose.

In the summer months of 1983 the building team managed to finish making and placing the two bottom boards, the ones closest to the keel.

The ship-builders were forced to make a number of primitive wooden clamps to keep the boards in place and in the correct shape until the floors were in place.

All the floors are in place. They keep the shape of the ship, and the workers have begun to make fast the very solid keelson, a very important part of a Viking ship. It is here the power from the ship's sail is transmitted through the mast to the hull of the ship. The keelson is made from one big, solid piece of oak wood.

It began to look like a Viking ship

Work on the ship continued throughout the winter of 1983-84, which luckily was one of the milder ones in this part of the world. The third, fourth and fifth boards were laid, and at the same time members of the ship-building-team again took to the forest in order to find crooked oak wood for floors, frames, knees and other parts of the ship which have an irregular shape.

The advantage of using naturally grown crooked timber that, roughly, has the shape of the required part, is that you can cut or split along the grain of the wood. In this way the finished part is both supple, strong and lightweight. This natural suppleness and strength in the seemingly slight material is, by the way, one of the most characteristic features of the Viking shipbuilding art.

One of the building-team using a homemade spoonbill-auger.

The keelson is made secure with an apparently simple, but in fact extremely refined system of knees and locks and the natural upward-growing branch which is fastened to two solid crossbeams in the middle of the ship.

The bow of the ship is trimmed with a chisel.

It takes both time and hard work to plane the long board-planks down to the required size.

The broadbladed axe is one of the classical ship-builder's tools.

Ingenuity was essential. It proved to be difficult to keep the long boards together in the correct curvature, but then the building-team discovered that part of the correct bend could be given in advance in this simple way.

While hewing out and trimming boards as well as other parts of the ship, the builders found out that the axes and other tools from the Viking Age were considerably better suited to the work than modern manufactured tools. Also in this field the experience of centuries has developed technically perfect things.

By February 1984 the work had progressed so far that it was possible to place floors in "Roar Ege," and thanks to the growing experience and proficiency of the builders, the work went ahead quickly.

The ship had, so to speak, "taught" the builders how things ought to be done.

The inside of the top board is reinforced with a "stringer," a long piece of wood fastened with wooden spikes.

Faster than ever

The summer of 1984 – right up to the culmination of the project, the launching of the ship – passed by quickly at the harbour in Roskilde.

A lot of work was done on the "Roar Ege," that already looked like a Viking ship. Many curious bystanders came to watch day by day, but for a time, however, interest was focussed on something else.

The rigging – hemp and flax

The rigging, or cordage, of the ship of course had to be made as authentic as humanly possible. A Viking ship with modern nylon rigging would not meet the strict scientific requirements of the whole project.

So, for this reason, both the running and the standing rigging for "Roar Ege" has been made in the oldfashioned way – by hand and by methods that the Vikings themselves may well have used.

The running rigging is the term for all the ropes used for hoisting sail and for steering the ship. The standing rigging, on the other hand, consists of the ropes that are stretched from the top of the mast to both bow,

The ropewalk on the lawn in front of the Viking Ship Museum was nearly one hundred metres long.

stern and the sides of the ship. These ropes simply have the function of keeping the mast firmly in place.

In June 1984 "Fur Rebslåeri" – a firm specializing in this age old technique – set up its portable ropewalk on the lawn in front of the Viking Ship Museum.

After a long and time-consuming process, hundreds of metres of twisted ropes, of seventy percent Russian hemp and thirty percent flax, were made.

A length of rope is nearly finished and wound on a wooden spool.

And finally the rope is dipped in wood-tar.

The top helps to keep the threads separate before they are twisted together to make a rope.

A "toggle," which is used in the rigging to fasten the rope to straps on the gunwale or to connect two rope ends. This also applies to the sail. A "toggle" is easy to release. More than three hundred "toggles" of different sizes and shapes were found in Hedeby. The shape and size of a "toggle" depends on where in the rigging it is used.

The bottom corner of the sail is held by sheets which, in the stern, pass through triangular holes placed high up on each side of the ship. These holes are a genial invention as several ropes can pass through them without hampering one another. When the sail is lowered the holes can be used for oars.

The stay is the rope in the standing rigging which goes from the top of the mast to the bow of the ship. Many Viking-ships not only had a bowstay, but also a stay in the stern of the ship.

Block for the middle sheet right under the sail.

Left: A block with wheel, reconstructed after a find at Hedeby. A block gives interchange like a gear so that less muscular strength is needed to hoist the sail. Right: A fender.

The square rig

The sail for "Roar Ege" was a special problem, not least because the experts at the Viking Ship Museum had not even the smallest remnant of an original sail to guide them.

Calculation, however, did much to help. The overall dimensions of the ship gave clear indications as to how high and how broad the sail should be. Historical knowledge in other fields also gives a reasonably firm basis for the assumption that Viking ships had sails of flax, homespun wool or hemp.

The first sail for "Roar Ege" was made of Scottish flax and at a later date the ship will also be fitted out with a sail made of wool, which is somewhat lighter than the heavy flax.

A woollen sail of that type must, by the way, be impregnated in a special way. It will be greased with a mixture of a decoction of birch-bark and horsefat. This ancient grease makes the sail wind and watertight.

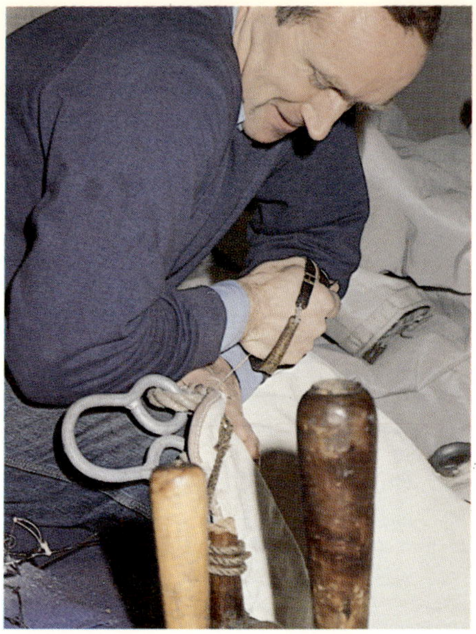

The top picture shows the sailmaker at work on "Roar Ege's" first sail, the one made of Scottish flax.
The bottom picture shows an ancient type loom that may well have been used by the Vikings for sail-making.

The rudder

A Viking ship has the rudder on its side, not in the stern. This, in our eyes, distinctive construction proves on closer examination to be eminently suited to its purpose. It is yet another proof of the Vikings' very advanced shipbuilding techniques.

But for the builders of "Roar Ege" the shape as well as the details of the rudder was a problem.

Scientists in the Viking Ship Museum had done an impressive amount of preliminary work. The rudders of the Gokstad, Oseberg and Ladby ships, the only well preserved ones of which we know, had been measured from end to end and the measurement figures subjected to all sorts of mathematical calculations. Many sketches of possible solutions had been made and it had gradually become clear how the rudder for the "Roar Ege" should be made.

The rudder passing through the water in calm sea.

The rudder-stock is on a "nipple" which keeps the rudder from the side of the ship, and consequently gives it the correct position in the water.

The helmsman does not use much strength to steer the ship – a situation from the sailing tests on Roskilde fiord.

The calculations proved that nothing in a Viking ship rudder is coincidential. The overall shape and each detail are very closely related to the shape and size of both the hull and sail of the ship.

The side rudder requires a perfect balance between sail and hull. A Viking ship can be steered with the sail alone. The rudder is there only for small course corrections.

In the "Roar Ege" it works perfectly.

The launching

When the "Roar Ege" was launched from the building site into the very water where her predecessor had sailed nearly a thousand years ago, the weather was perfect: a bright, sunny and clear autumn day.

From the early hours in the morning the builders were busy indeed. Many last minute

The "Roar Ege" has at last been given its name. The ship is named after a legendary Danish king of the Scyldings family. Moreover, some sources name King Roar as the founder of Roskilde, so the name of the ship could hardly be more suitably chosen!

When the ten metre high mast was hoisted, some of the building-team had to haul on the ropes.

There were mountains of sandwiches for the guests at the launching ceremony.

A practical arrangement (naturally there was fiddle music on this great ocasion) the launching.

details had to be worked out before the solemn christening ceremony and the exiting moment of launching.

"Roar Ege" was christened with water from the most ancient spring in Roskilde, dating back almost to the Viking Age.

The launching was witnessed by nearly one thousand spectators and film and television crews from several countries recorded the event for posterity.

Finally, the members of the building team placed themselves on each side of the ship and partly carried, partly hauled her into the water.

The man who started the whole thing: Mr. Ole Crumlin Pederse,n ex-curator of the Viking Ship Museum, was responsible for the practical management of the launching.

Hundreds of spectators watched the festive launching of the "Roar Ege."

Launched and then what?

The "Roar Ege" was now floating on the waters of Roskilde fiord, where its ancestor had sailed a thousand years ago.

Everybody was happy. The members of the building team performed strange dances on the deck of the new ship. Two years of hard often tough work had come to a successful end.

And now what?

First of all a holiday trip for the builders. A couple of weeks cruise in Zealand waters.

But then the serious part began. "Roar Ege" had to be tested in trial runs under scientific control with the most modern equipment.

Electronic navigational gear was installed and the ship left Roskilde harbour to re-examine all the theories on an exact copy of a Viking ship.

For the first sailing-trails very modern navigational equipment was temporarily taken on board.

The ballast – several tons of stones – keep the "Roar Ege" on an even keel even when the wind is freshening.

The first decisive conclusions have already been reached. They confirm in every respect all the hopes and expectations with regard to seaworthiness and sailing ability. "Roar Ege" is not especially fast, but moves beautifully in choppy water.

She is - and the original, "Skuldelev No. 3" was - a technically superb prototype of a small trading vessel from the Viking Age, developed through centuries for the purpose of beating around in fiords yet big and seaworthy enough to be able to go to Norway or to England.

She must have been very well suited to the needs of the Viking trader. With a crew of five or six onboard, the trader could relax while the ship almost sailed itself.

The superb qualities of the square-sail rig is also once more reaffirmed. Our Viking ancestors could beat to windward, sail close to the wind and navigate precisely in narrow waters.

"Roar Ege" could of course also be rowed. There is room for two pairs of oars.

In the final summing up, it must be clearly stated, that "Roar Ege" is an unqualified scientific and technical triumph, not only for the original Viking shipbuilders, but also for those who followed their example.

"Roar Ege's" long yard is secured to the mast with a "parrel," a curved piece of wood which must be greased in order to slide easily up and down the mast.

The square-sail lugs well in the fresh wind on Roskilde fiord.

"Roar Ege's" data:

Overall length: 14.2 metres
Length (in water-line): 10.7 metres
Width: 3.3 metres
Sail: 46 square metres
Ballast: 3.7 metric tonne
Total weight during tests (displacement): 6.0 tons
Man hours: 15,000
Consumed oak and fir: 25 metric tonne
Weight without ballast (with mast and yard): 2.45 metric tonne

Far beyond the horizon

The Vikings sailed far indeed in their amazing ships. This map shows us how far, both east and west through Russian rivers to the Black Sea and the Caspian Sea, west across the North Sea and the Atlantic to Greenland – and America.